City of Tomorrow

by Stacie Leto © God, 2015

(by Stacie Marie van Nielen)

With special thanks to:

Jared Joseph Leto

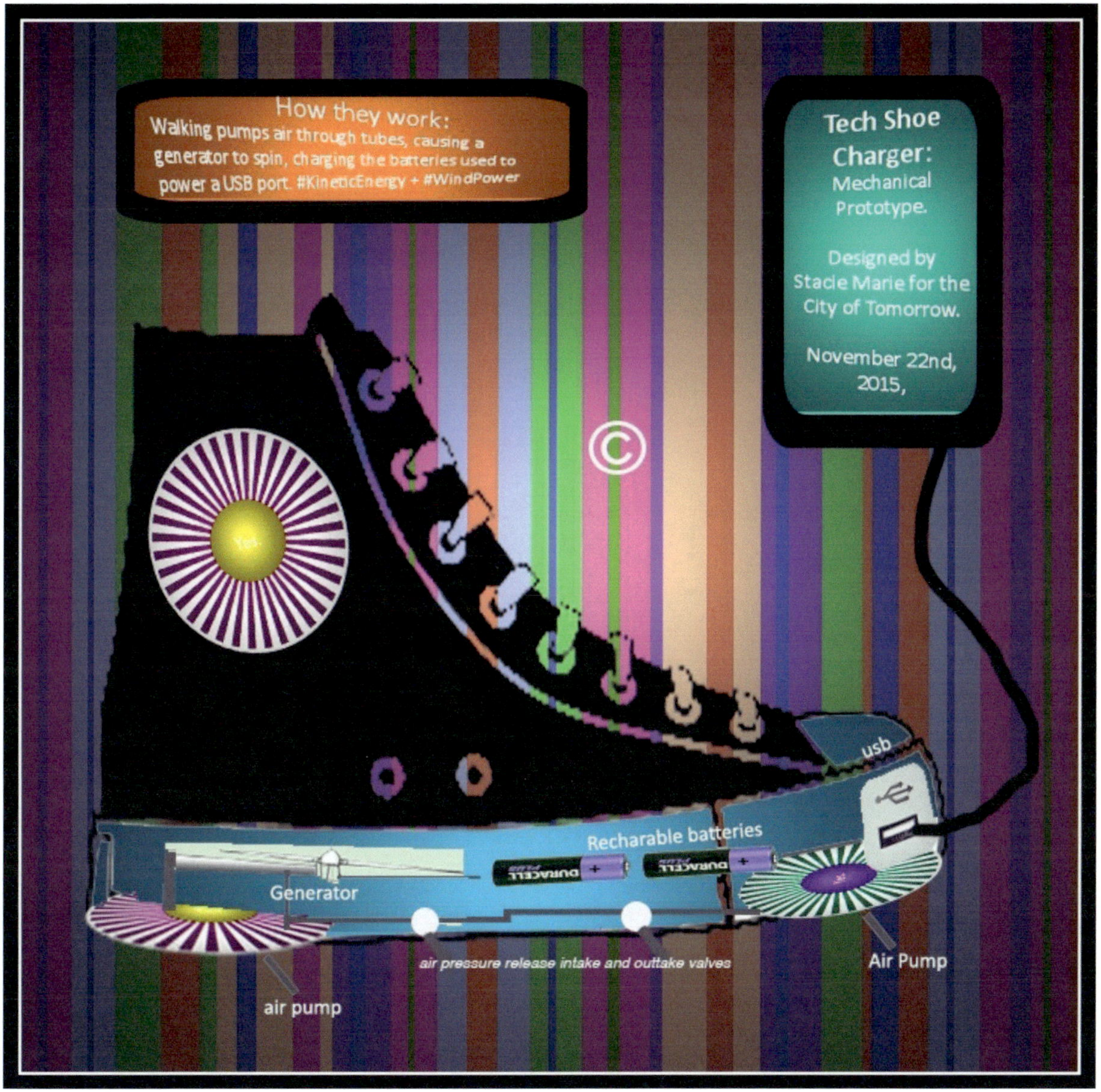
How they work:
Walking pumps air through tubes, causing a generator to spin, charging the batteries used to power a USB port. #KineticEnergy + #WindPower
Tech Shoe Charger:
Mechanical Prototype.
Designed by Stacie Marie for the City of Tomorrow.
November 22nd, 2015,
usb
Recharable batteries
Generator
air pressure release intake and outtake valves
Air Pump
air pump

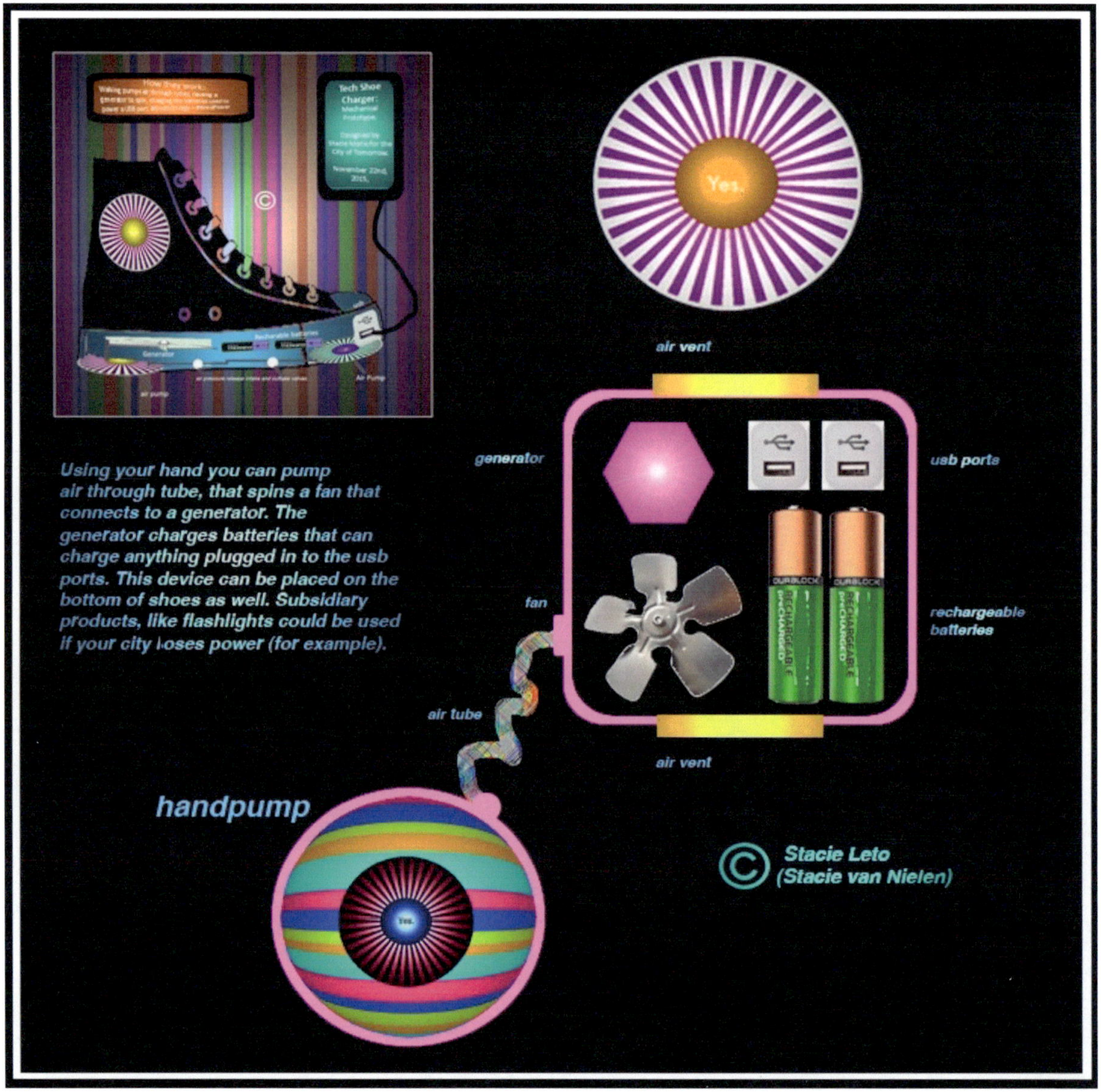
Tech Shoe
Charger:
November 22nd,
2015,
Generator
Air Pump
air pump
Using your hand you can pump
air through tube, that spins a fan that
connects to a generator. The
generator charges batteries that can
charge anything plugged in to the usb
ports. This device can be placed on the
bottom of shoes as well. Subsidiary
products, like flashlights could be used
if your city loses power (for example).
Yes.
air vent
generator
usb ports
fan
rechargeable
batteries
air tube
air vent
handpump
Yes.
© Stacie Leto
(Stacie van Nielen)

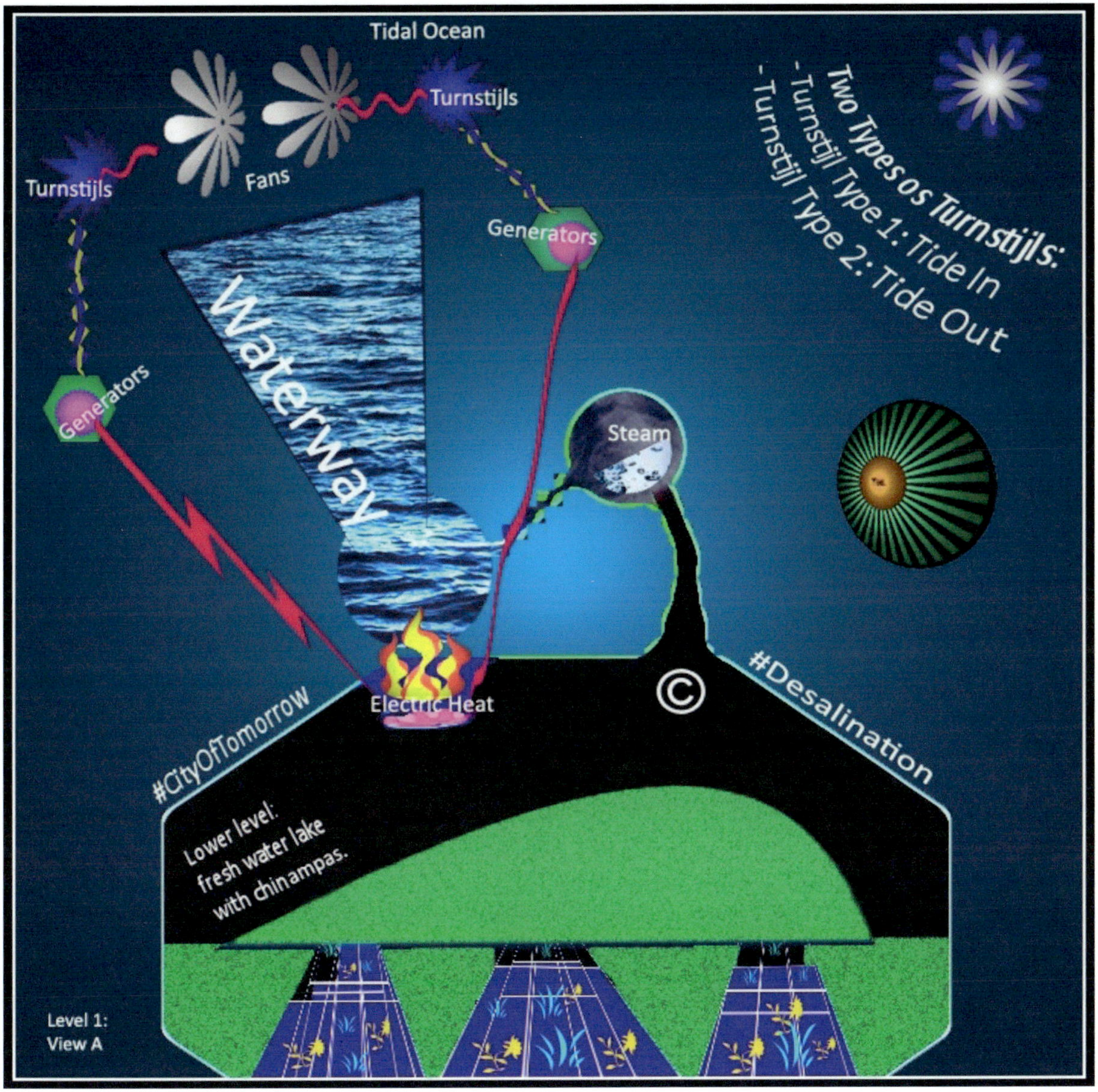
Tidal Ocean
Turnstijls
Fans
Turnstijls
Two Types os Turnstijls:
- Turnstijl Type 1: Tide In
- Turnstijl Type 2: Tide Out
Generators
Generators
Waterway
Steam
Electric Heat
©
#CityOfTomorrow
#Desalination
Lower level:
fresh water lake
with chinampas.
Level 1:
View A

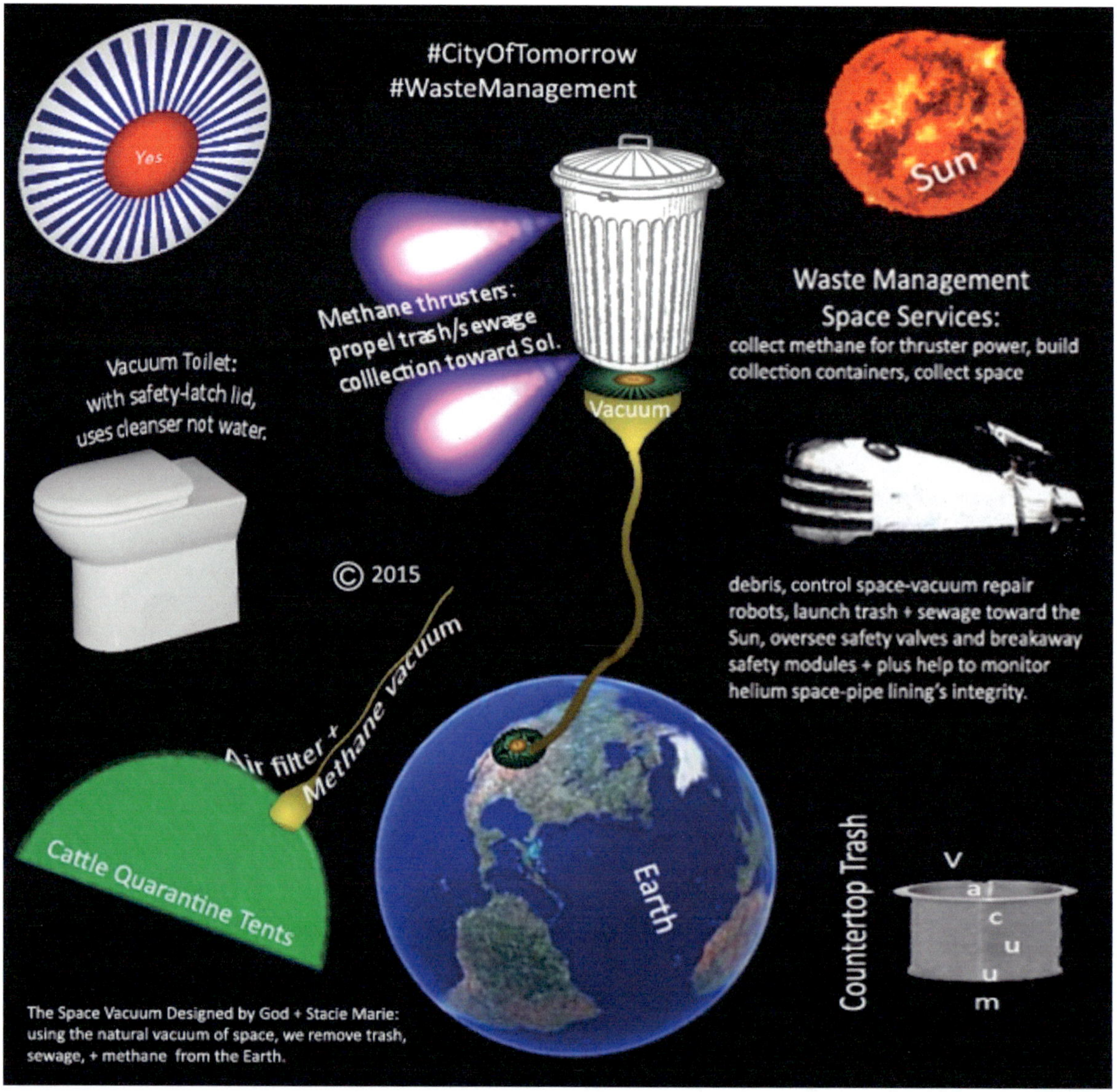
#CityOfTomorrow
#WasteManagement
Yes
Sun
Methane thrusters: propel trash/sewage colllection toward Sol.
Vacuum Toilet: with safety-latch lid, uses cleanser not water.
Vacuum
Waste Management Space Services:
collect methane for thruster power, build collection containers, collect space
debris, control space-vacuum repair robots, launch trash + sewage toward the Sun, oversee safety valves and breakaway safety modules + plus help to monitor helium space-pipe lining's integrity.
© 2015
Air filter + Methane vacuum
Cattle Quarantine Tents
Earth
Countertop Trash
Vacuum
The Space Vacuum Designed by God + Stacie Marie: using the natural vacuum of space, we remove trash, sewage, + methane from the Earth.

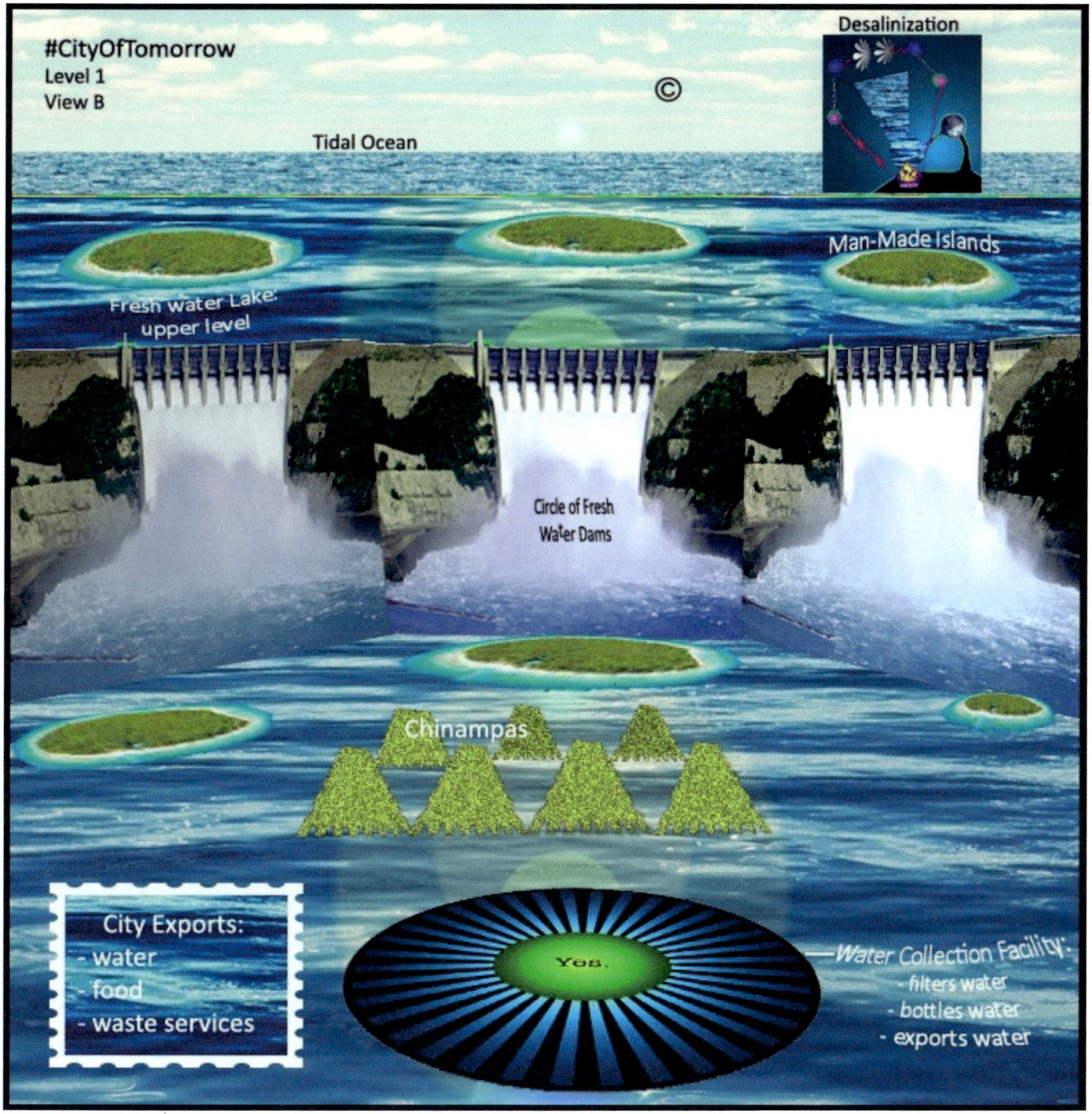
#CityOfTomorrow
Level 1
View B
©
Tidal Ocean
Desalinization
Man-Made Islands
Fresh water Lake:
upper level
Circle of Fresh
Water Dams
Chinampas
City Exports:
- water
- food
- waste services
Yes.
Water Collection Facility:
- filters water
- bottles water
- exports water

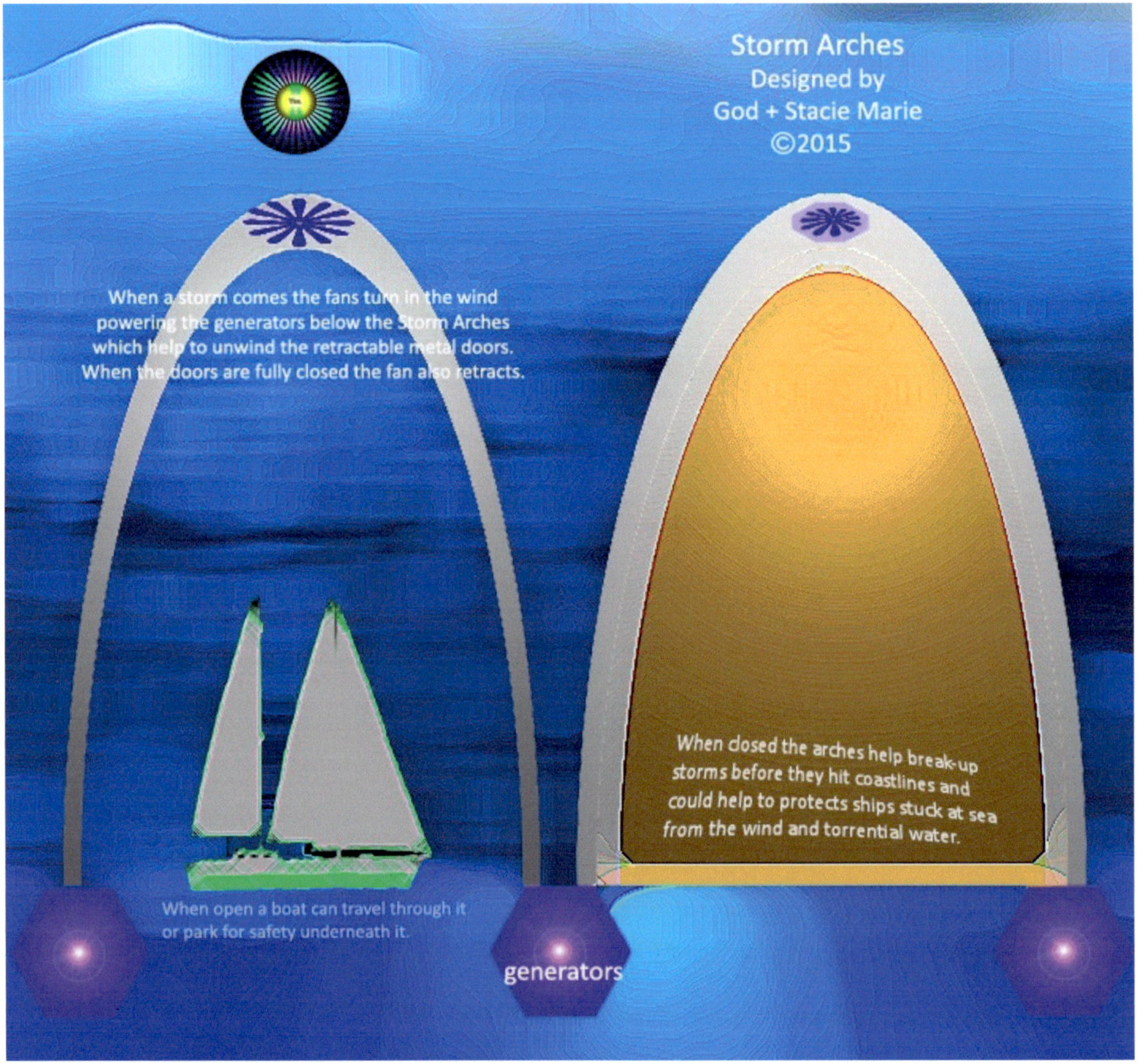
Storm Arches
Designed by
God + Stacie Marie
©2015
When a storm comes the fans turn in the wind
powering the generators below the Storm Arches
which help to unwind the retractable metal doors.
When the doors are fully closed the fan also retracts.
When closed the arches help break-up
storms before they hit coastlines and
could help to protects ships stuck at sea
from the wind and torrential water.
When open a boat can travel through it
or park for safety underneath it.
generators

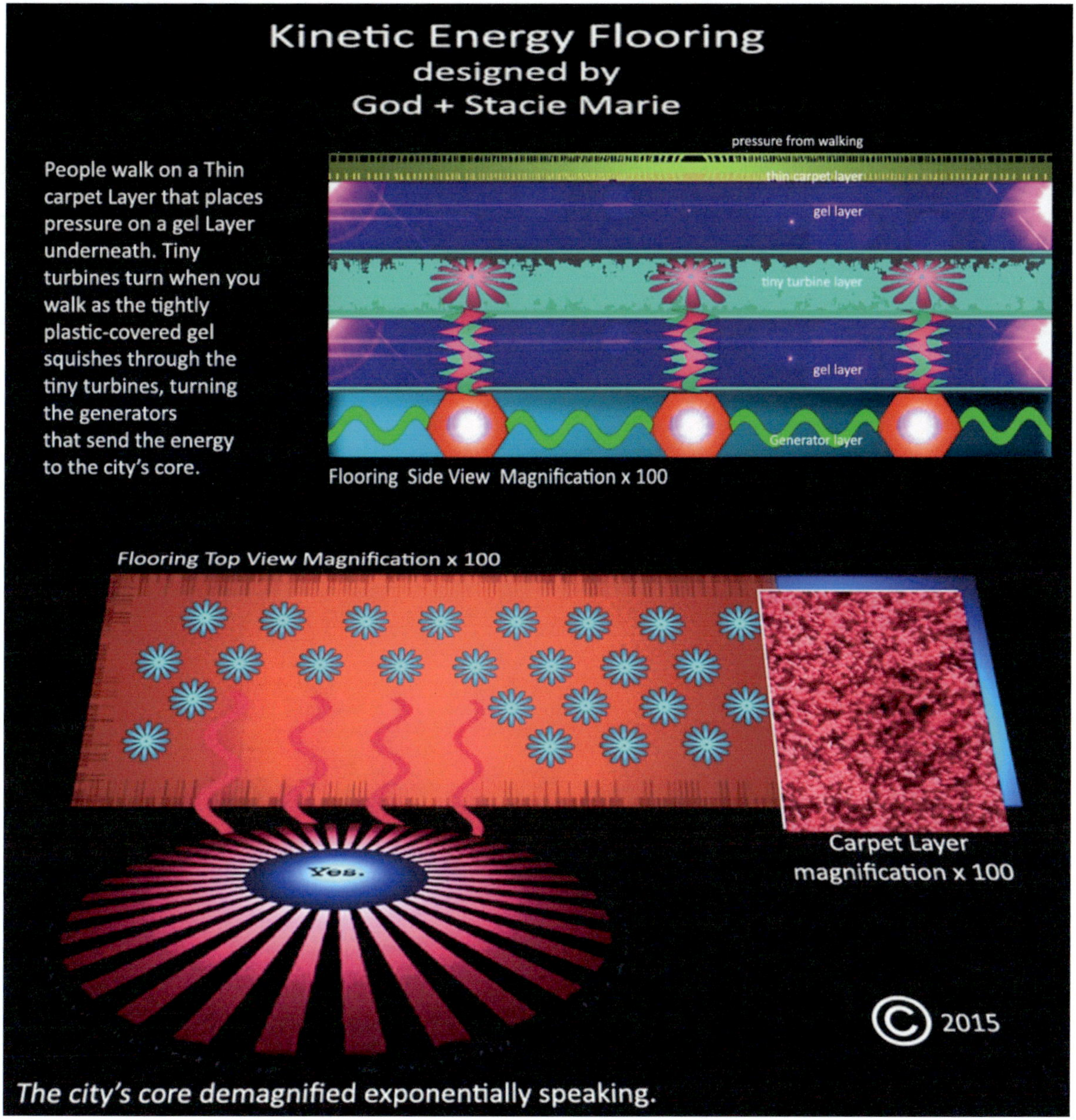
Kinetic Energy Flooring
designed by
God + Stacie Marie
People walk on a Thin carpet Layer that places pressure on a gel Layer underneath. Tiny turbines turn when you walk as the tightly plastic-covered gel squishes through the tiny turbines, turning the generators that send the energy to the city's core.
pressure from walking
thin carpet layer
gel layer
tiny turbine layer
gel layer
Generator layer
Flooring Side View Magnification x 100
Flooring Top View Magnification x 100
Yes.
Carpet Layer
magnification x 100
© 2015
The city's core demagnified exponentially speaking.

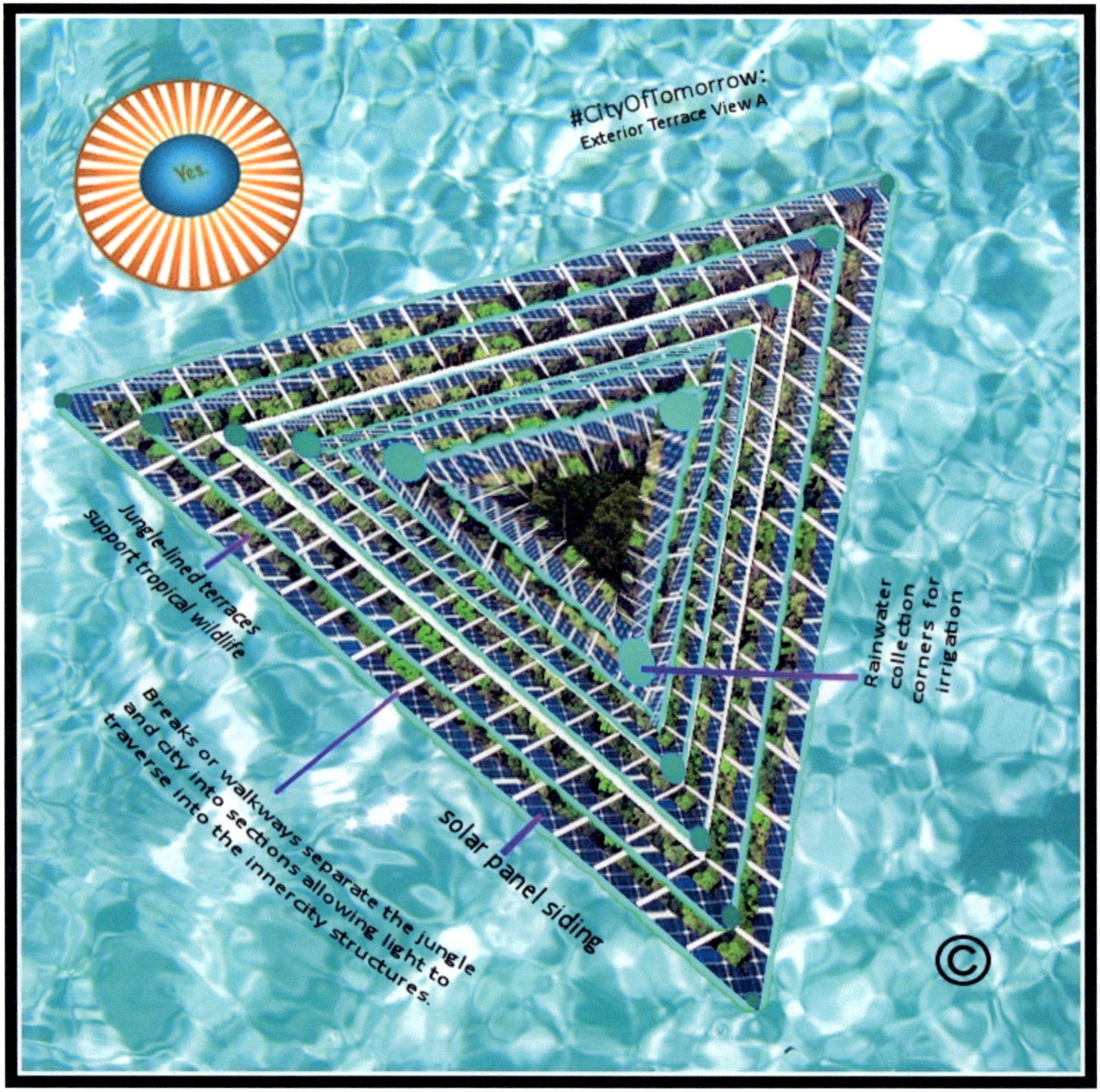

Yes
#CityOfTomorrow:
Exterior Terrace View A
Jungle-lined terraces
support tropical wildlife
Breaks or walkways separate the jungle
and city into sections allowing light to
traverse into the innercity structures.
solar panel siding
Rainwater
collection
corners for
irrigation
©

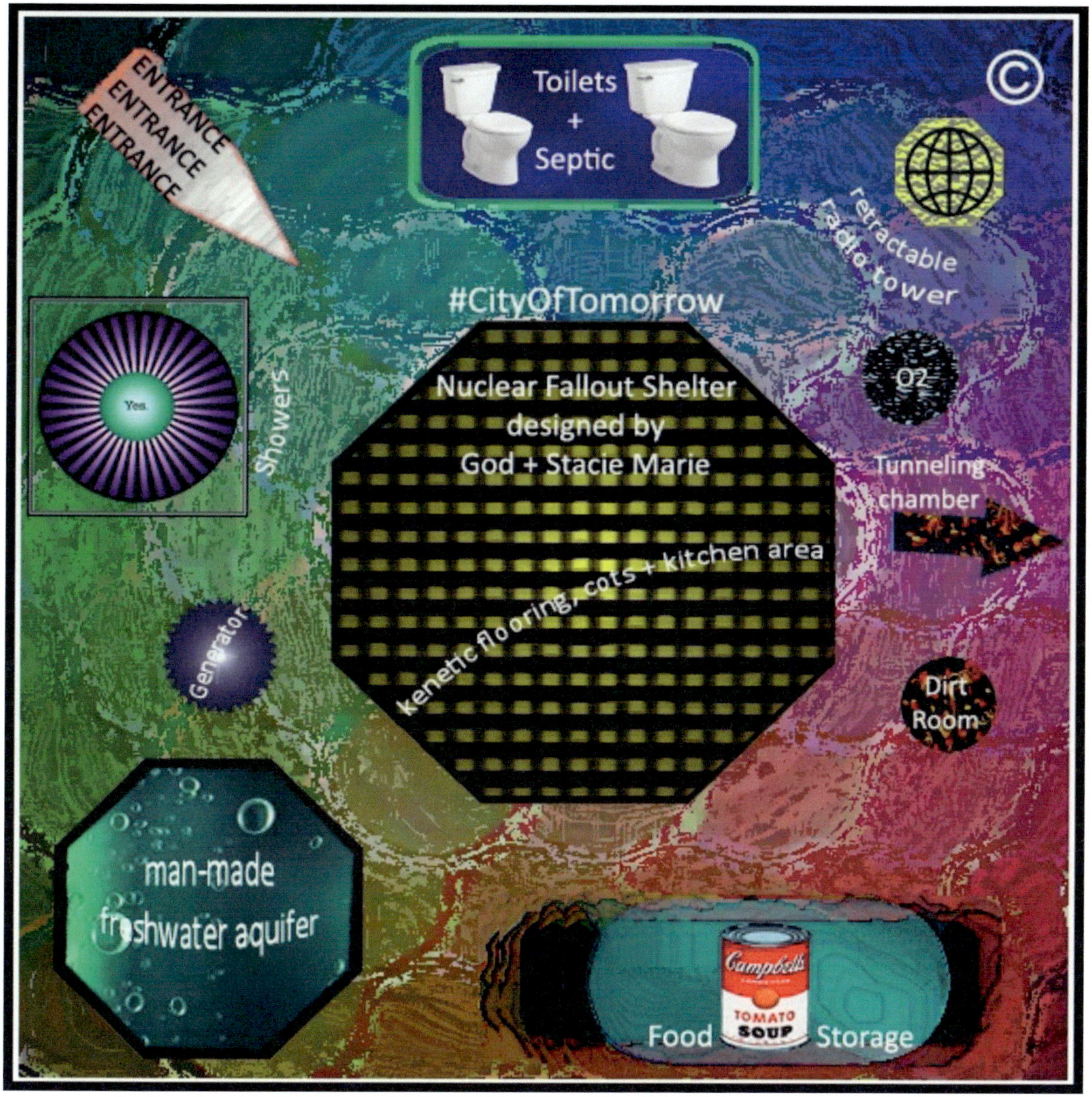

ENTRANCE
ENTRANCE
ENTRANCE
Toilets
+
Septic
©
retractable
radio-tower
#CityOfTomorrow
Yes.
Showers
Nuclear Fallout Shelter
designed by
God + Stacie Marie
O2
Tunneling
chamber
kenetic flooring, cots + kitchen area
Generator
Dirt
Room
man-made
freshwater aquifer
Campbell's
TOMATO
SOUP
Food
Storage

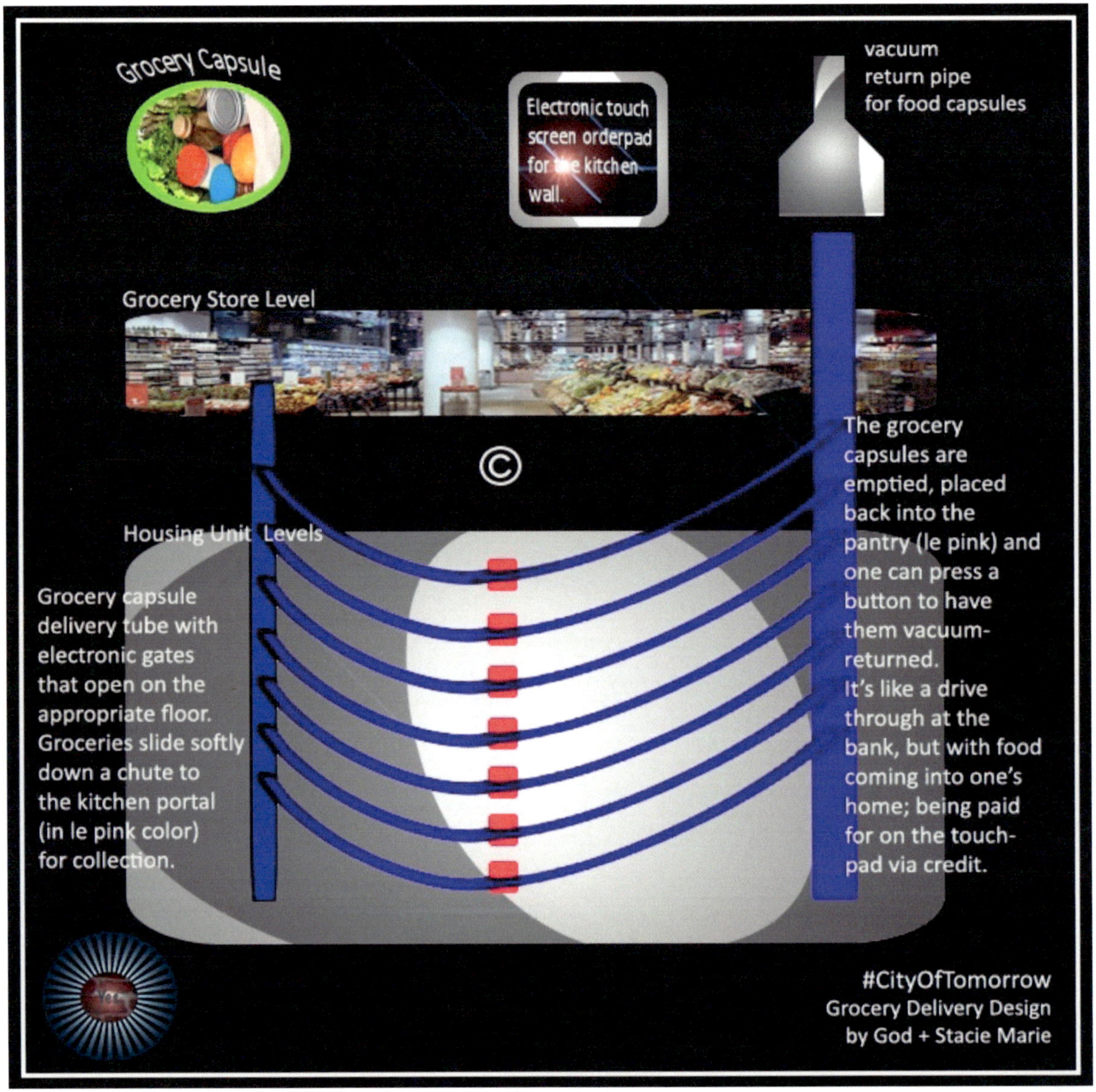
Grocery Capsule
Electronic touch screen orderpad for the kitchen wall.
vacuum return pipe for food capsules
Grocery Store Level
©
Housing Unit Levels
Grocery capsule delivery tube with electronic gates that open on the appropriate floor. Groceries slide softly down a chute to the kitchen portal (in le pink color) for collection.
The grocery capsules are emptied, placed back into the pantry (le pink) and one can press a button to have them vacuum-returned.
It's like a drive through at the bank, but with food coming into one's home; being paid for on the touch-pad via credit.
#CityOfTomorrow
Grocery Delivery Design
by God + Stacie Marie

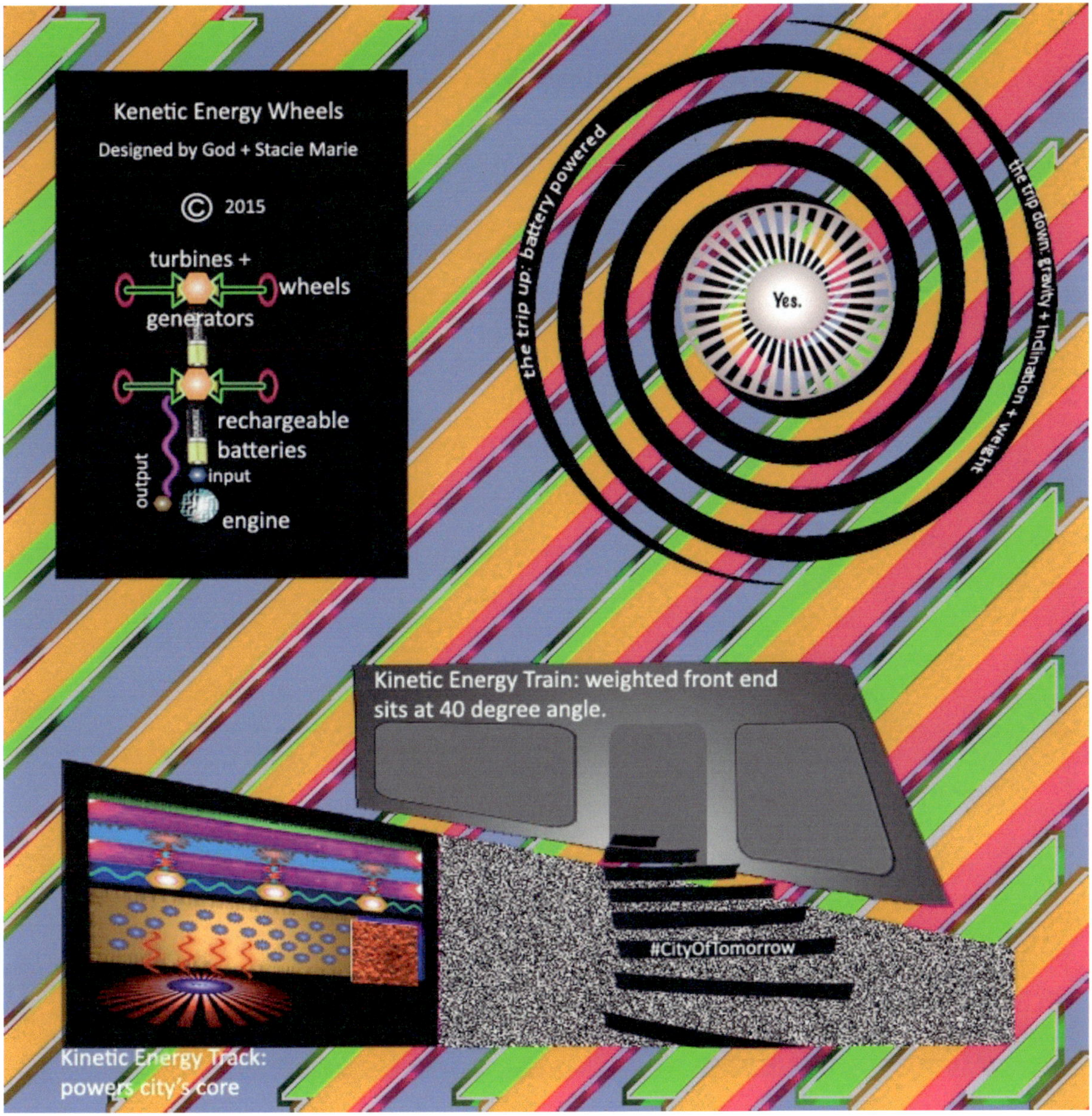
Kenetic Energy Wheels
Designed by God + Stacie Marie
© 2015
turbines +
wheels
generators
rechargeable
batteries
input
output
engine
the trip up: battery powered
the trip down: gravity + Inclination + weight
Yes.
Kinetic Energy Train: weighted front end sits at 40 degree angle.
#CityOfTomorrow
Kinetic Energy Track: powers city's core

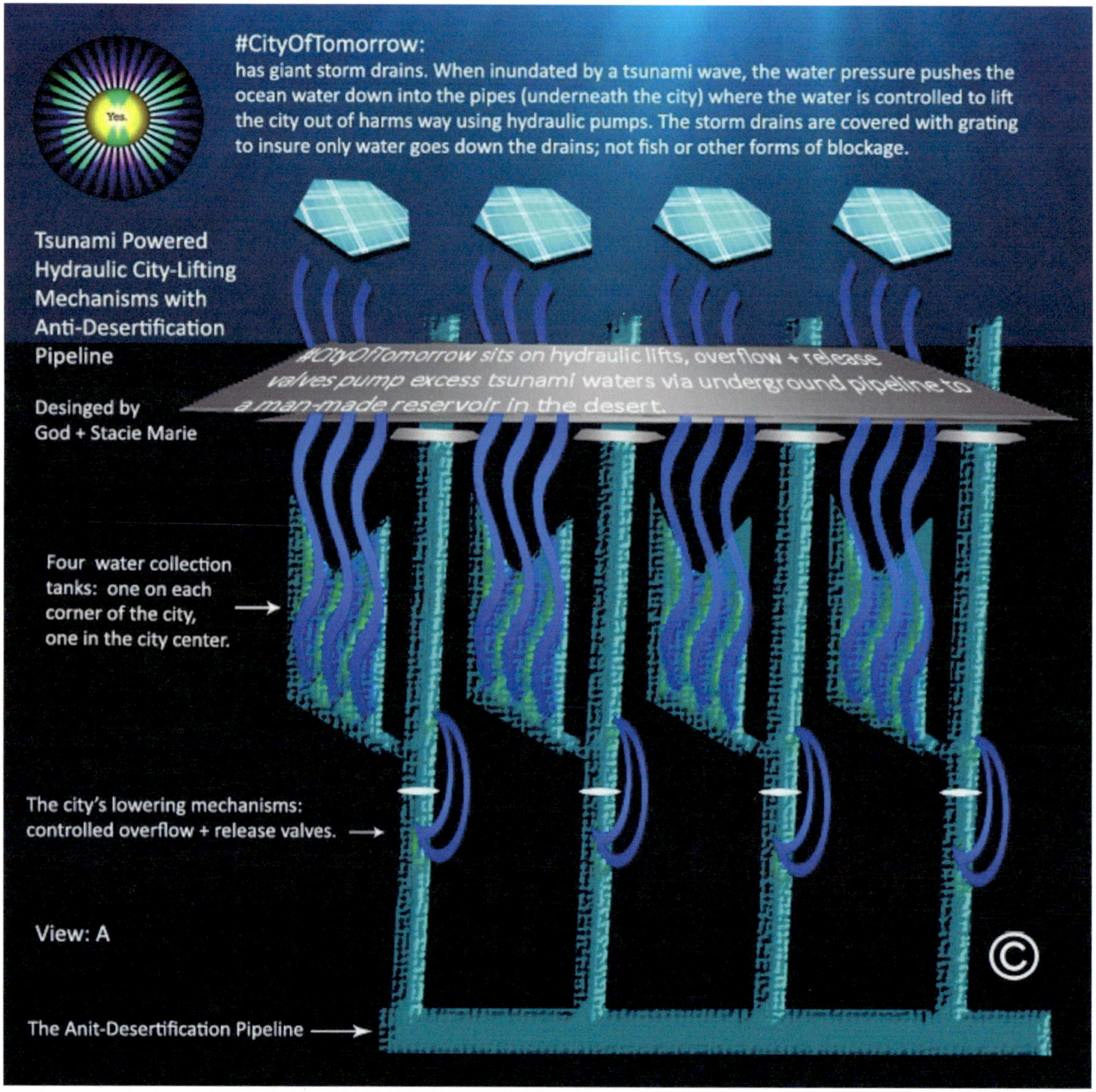
#CityOfTomorrow:
has giant storm drains. When inundated by a tsunami wave, the water pressure pushes the ocean water down into the pipes (underneath the city) where the water is controlled to lift the city out of harms way using hydraulic pumps. The storm drains are covered with grating to insure only water goes down the drains; not fish or other forms of blockage.
Yes.
Tsunami Powered Hydraulic City-Lifting Mechanisms with Anti-Desertification Pipeline
#CityOfTomorrow sits on hydraulic lifts, overflow + release valves pump excess tsunami waters via underground pipeline to a man-made reservoir in the desert.
Desinged by God + Stacie Marie
Four water collection tanks: one on each corner of the city, one in the city center.
The city's lowering mechanisms: controlled overflow + release valves.
View: A
©
The Anit-Desertification Pipeline

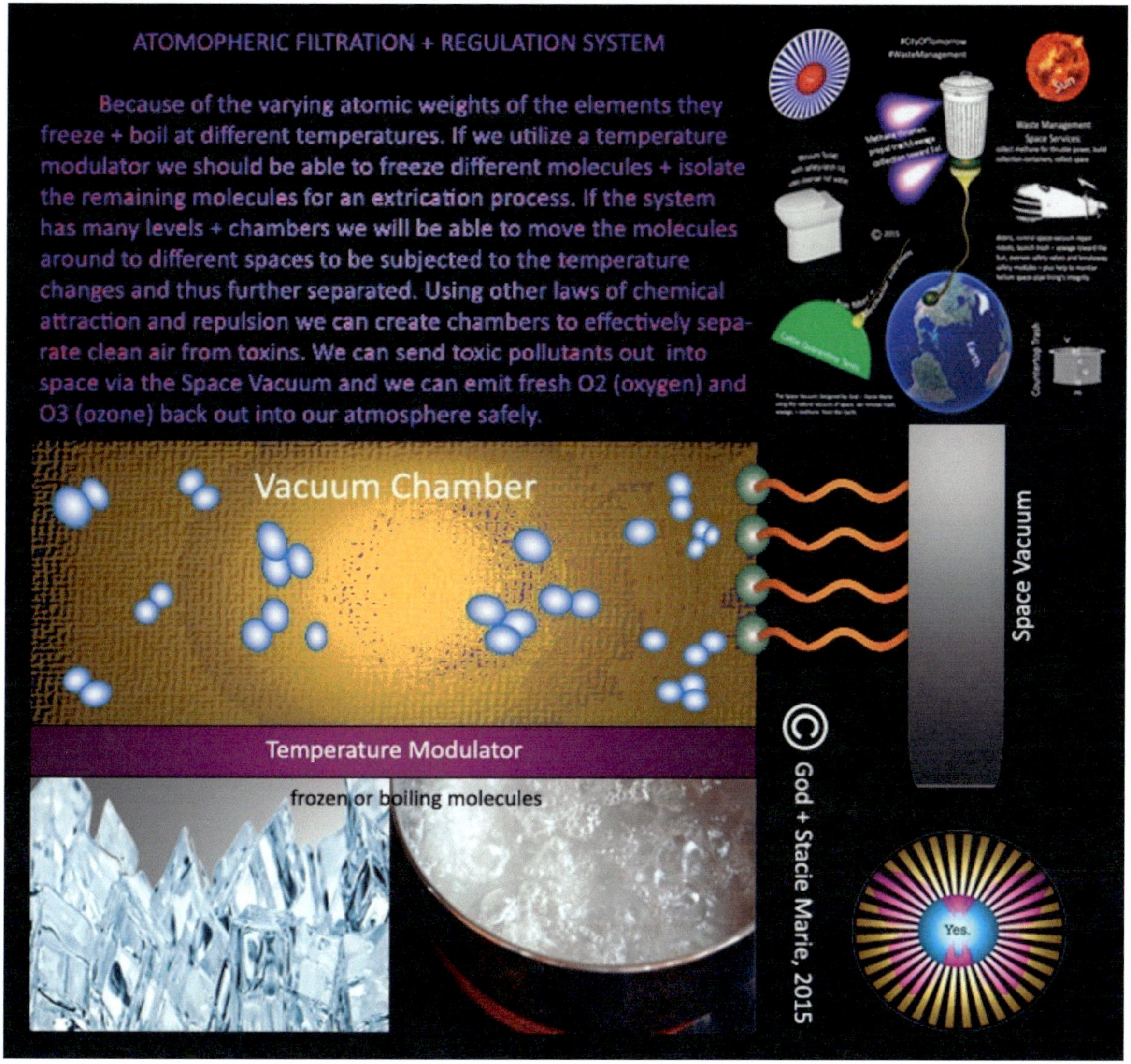
ATOMOPHERIC FILTRATION + REGULATION SYSTEM
Because of the varying atomic weights of the elements they freeze + boil at different temperatures. If we utilize a temperature modulator we should be able to freeze different molecules + isolate the remaining molecules for an extrication process. If the system has many levels + chambers we will be able to move the molecules around to different spaces to be subjected to the temperature changes and thus further separated. Using other laws of chemical attraction and repulsion we can create chambers to effectively separate clean air from toxins. We can send toxic pollutants out into space via the Space Vacuum and we can emit fresh O2 (oxygen) and O3 (ozone) back out into our atmosphere safely.
#CityOfTomorrow
#WasteManagement
Sun
Waste Management
Space Services:
Vacuum Chamber
Space Vacuum
Temperature Modulator
frozen or boiling molecules
© God + Stacie Marie, 2015
Yes.

All Safety-Bridges dedicated to the Legacy of President Obama
#CityOfTomorrow
©
Safety-Bridge Plaques
Bridges with Safety Slides

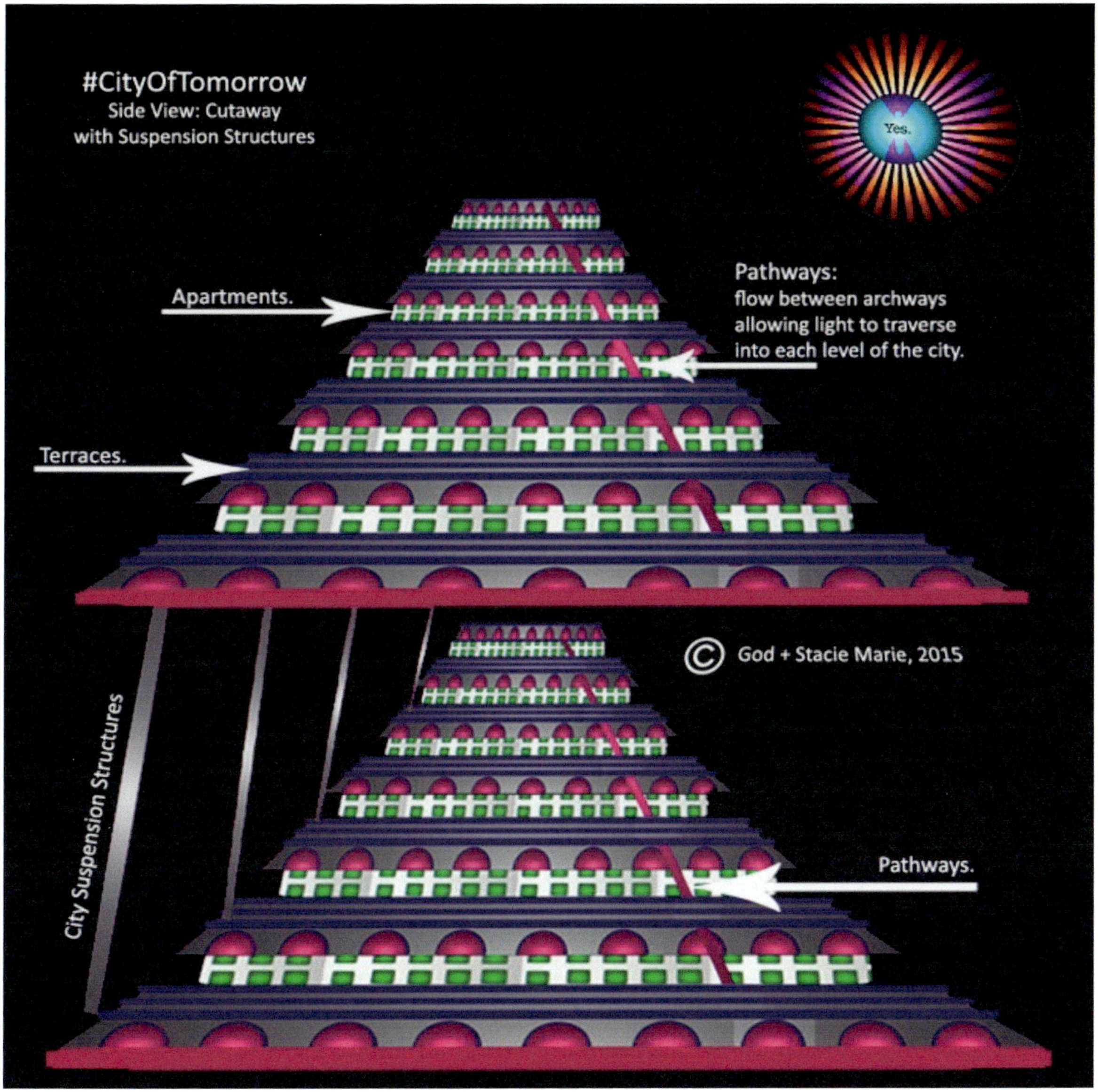
#CityOfTomorrow
Side View: Cutaway
with Suspension Structures
Yes.
Pathways:
flow between archways
allowing light to traverse
into each level of the city.
Apartments.
Terraces.
© God + Stacie Marie, 2015
City Suspension Structures
Pathways.

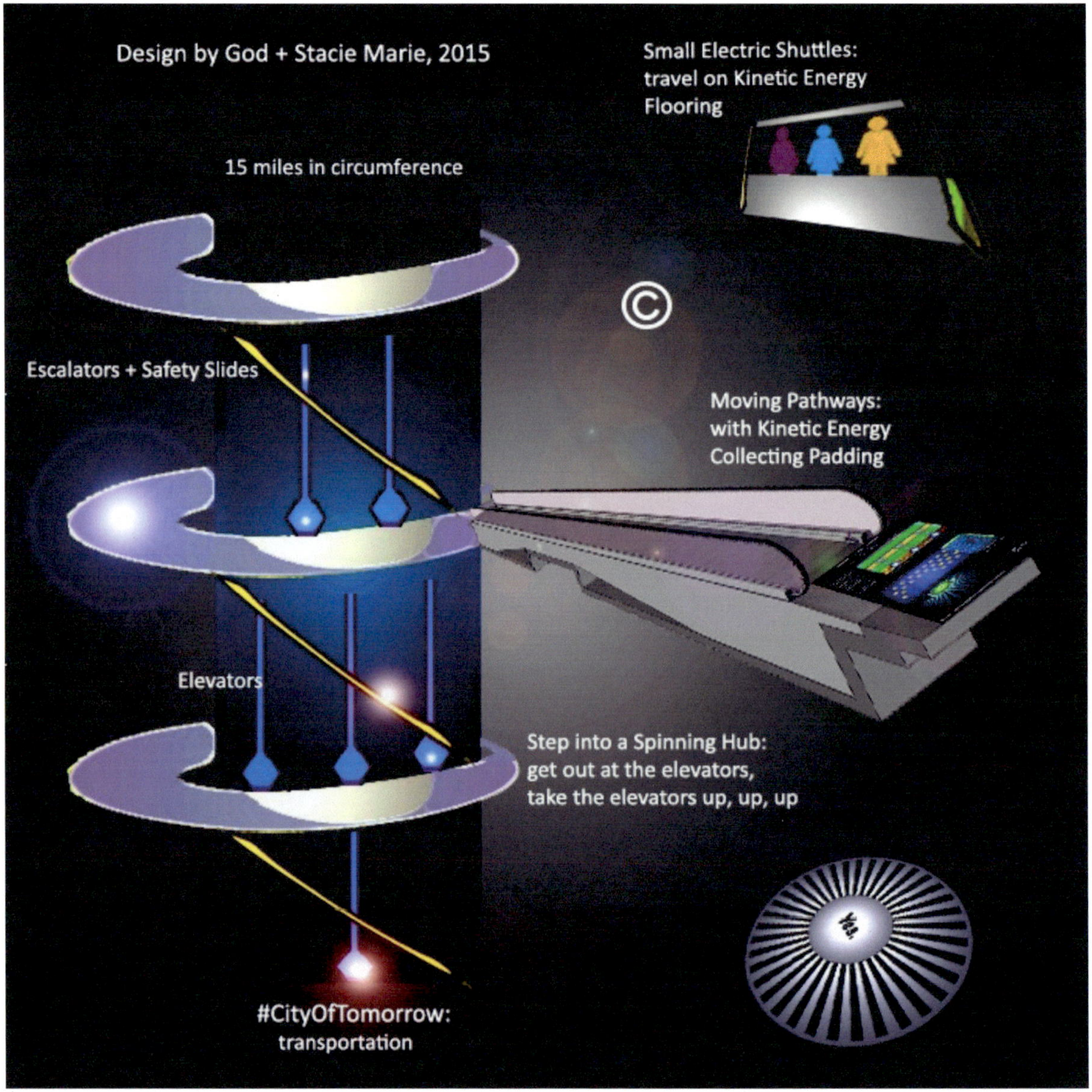
Design by God + Stacie Marie, 2015
Small Electric Shuttles:
travel on Kinetic Energy
Flooring
15 miles in circumference
©
Escalators + Safety Slides
Moving Pathways:
with Kinetic Energy
Collecting Padding
Elevators
Step into a Spinning Hub:
get out at the elevators,
take the elevators up, up, up
#CityOfTomorrow:
transportation

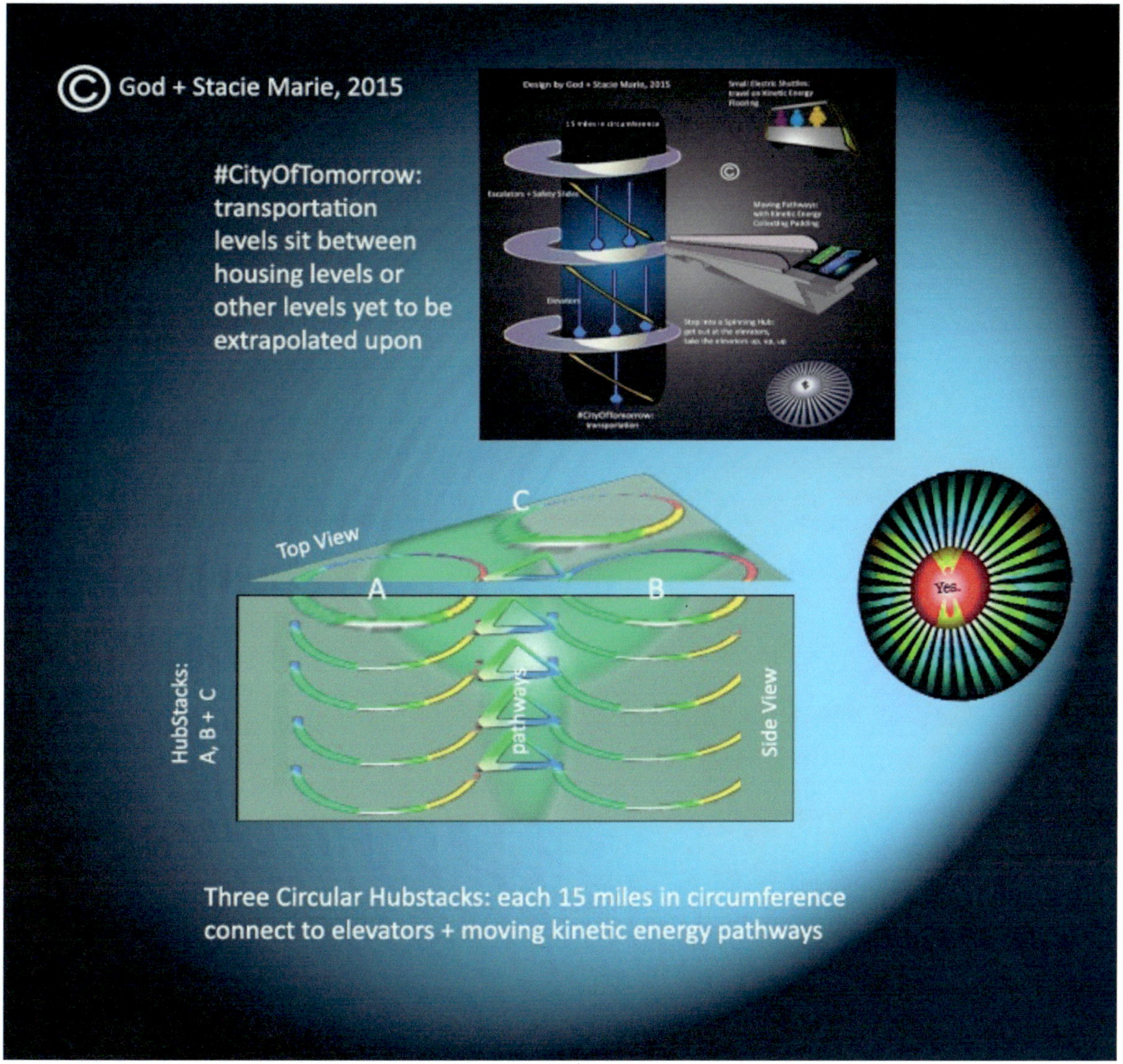

© God + Stacie Marie, 2015
#CityOfTomorrow:
transportation
levels sit between
housing levels or
other levels yet to be
extrapolated upon
Design by God + Stacie Marie, 2015
Top View
C
A
B
HubStacks:
A, B + C
pathways
Side View
Yes.
Three Circular Hubstacks: each 15 miles in circumference
connect to elevators + moving kinetic energy pathways

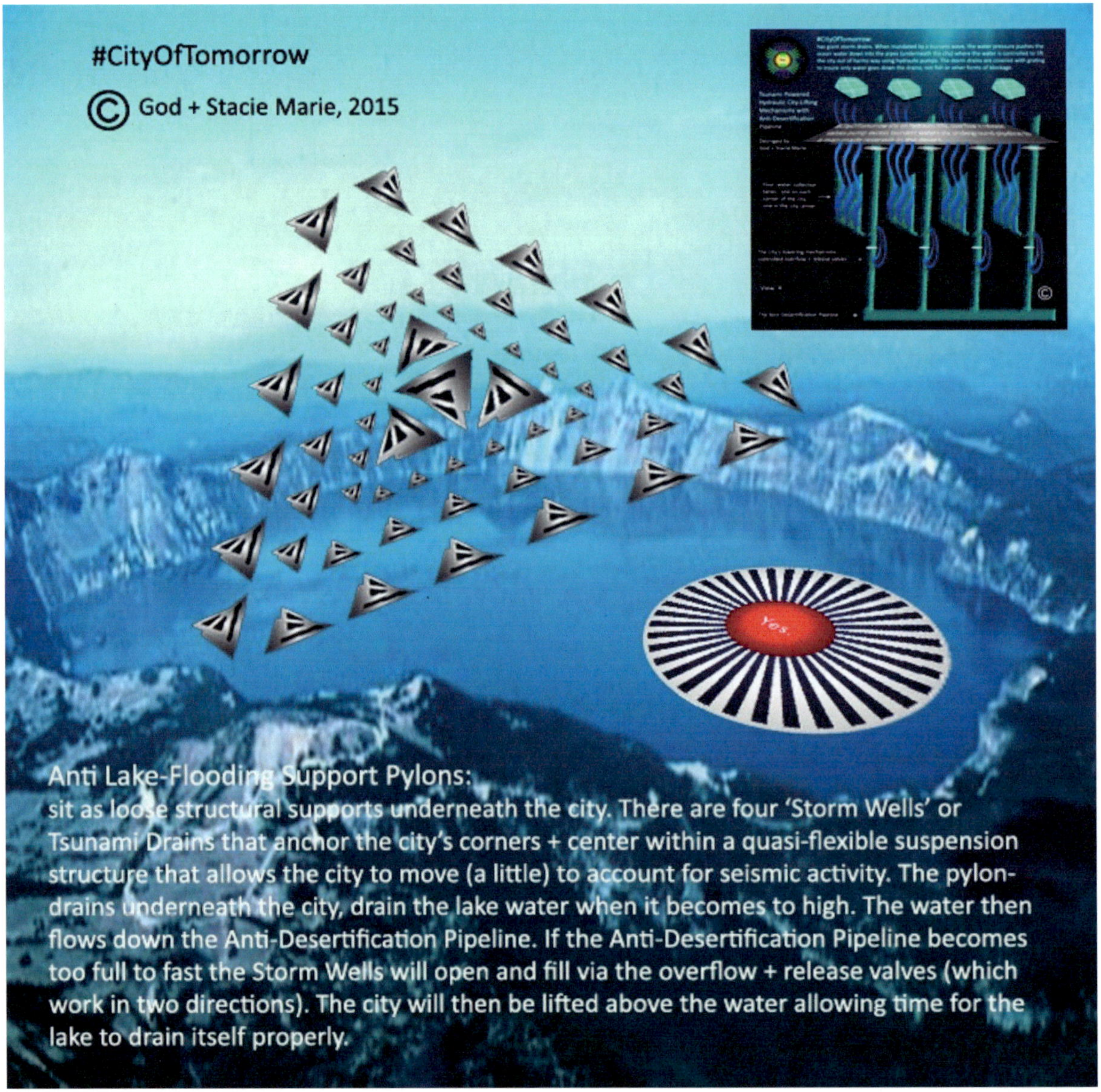
#CityOfTomorrow
© God + Stacie Marie, 2015
Yes
Anti Lake-Flooding Support Pylons:
sit as loose structural supports underneath the city. There are four 'Storm Wells' or Tsunami Drains that anchor the city's corners + center within a quasi-flexible suspension structure that allows the city to move (a little) to account for seismic activity. The pylon-drains underneath the city, drain the lake water when it becomes to high. The water then flows down the Anti-Desertification Pipeline. If the Anti-Desertification Pipeline becomes too full to fast the Storm Wells will open and fill via the overflow + release valves (which work in two directions). The city will then be lifted above the water allowing time for the lake to drain itself properly.

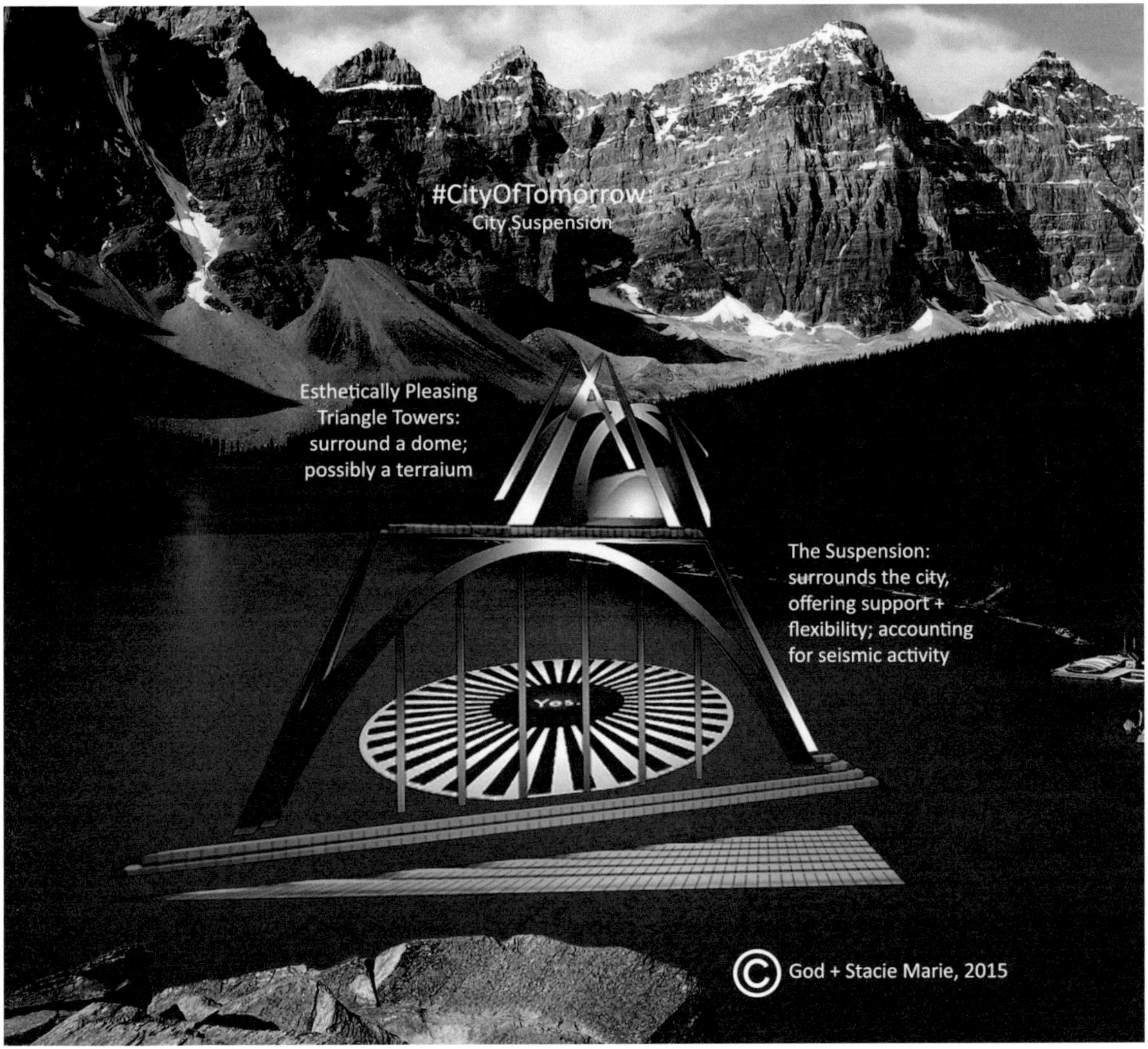
#CityOfTomorrow:
City Suspension
Esthetically Pleasing
Triangle Towers:
surround a dome;
possibly a terraium
The Suspension:
surrounds the city,
offering support +
flexibility; accounting
for seismic activity
Yes.
© God + Stacie Marie, 2015

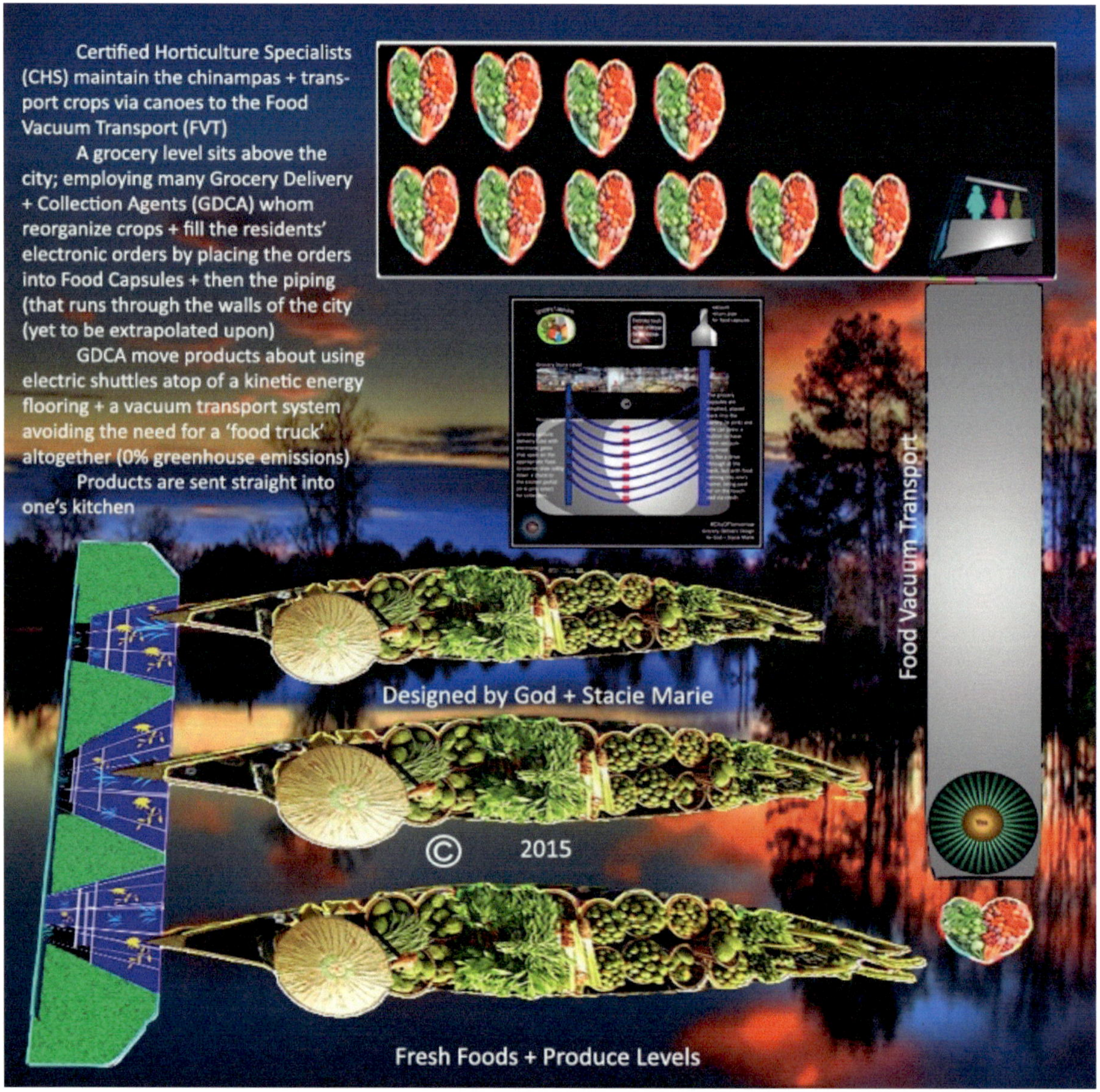
Certified Horticulture Specialists (CHS) maintain the chinampas + transport crops via canoes to the Food Vacuum Transport (FVT)
A grocery level sits above the city; employing many Grocery Delivery + Collection Agents (GDCA) whom reorganize crops + fill the residents' electronic orders by placing the orders into Food Capsules + then the piping (that runs through the walls of the city (yet to be extrapolated upon)
GDCA move products about using electric shuttles atop of a kinetic energy flooring + a vacuum transport system avoiding the need for a 'food truck' altogether (0% greenhouse emissions)
Products are sent straight into one's kitchen
Food Vacuum Transport
Designed by God + Stacie Marie
© 2015
Fresh Foods + Produce Levels

Pathway View: Artistic

Terraces with retaining walls sit above or outside of the apartment walls connected by pathways + bridges with Kinetic Energy Flooring designed to look like wooden bridges. There's a lovely jungle-adventure type of feel to it - like you are surrounded by nature when you traverse the housing structures

Designed by God + Stacie Marie, 2015

©

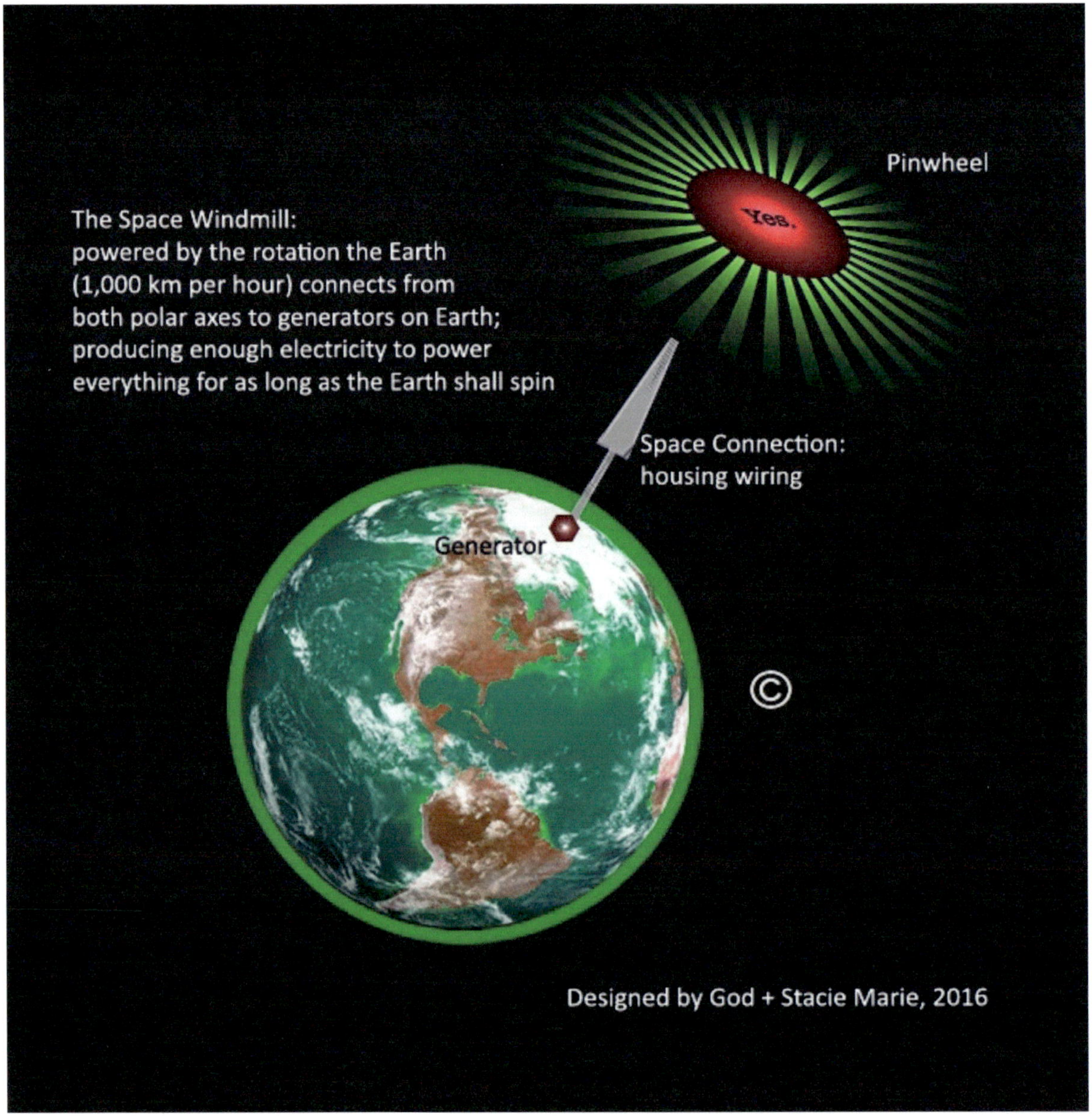

Pinwheel
Yes.
The Space Windmill:
powered by the rotation the Earth
(1,000 km per hour) connects from
both polar axes to generators on Earth;
producing enough electricity to power
everything for as long as the Earth shall spin
Space Connection:
housing wiring
Generator
©
Designed by God + Stacie Marie, 2016

Bonus Content:

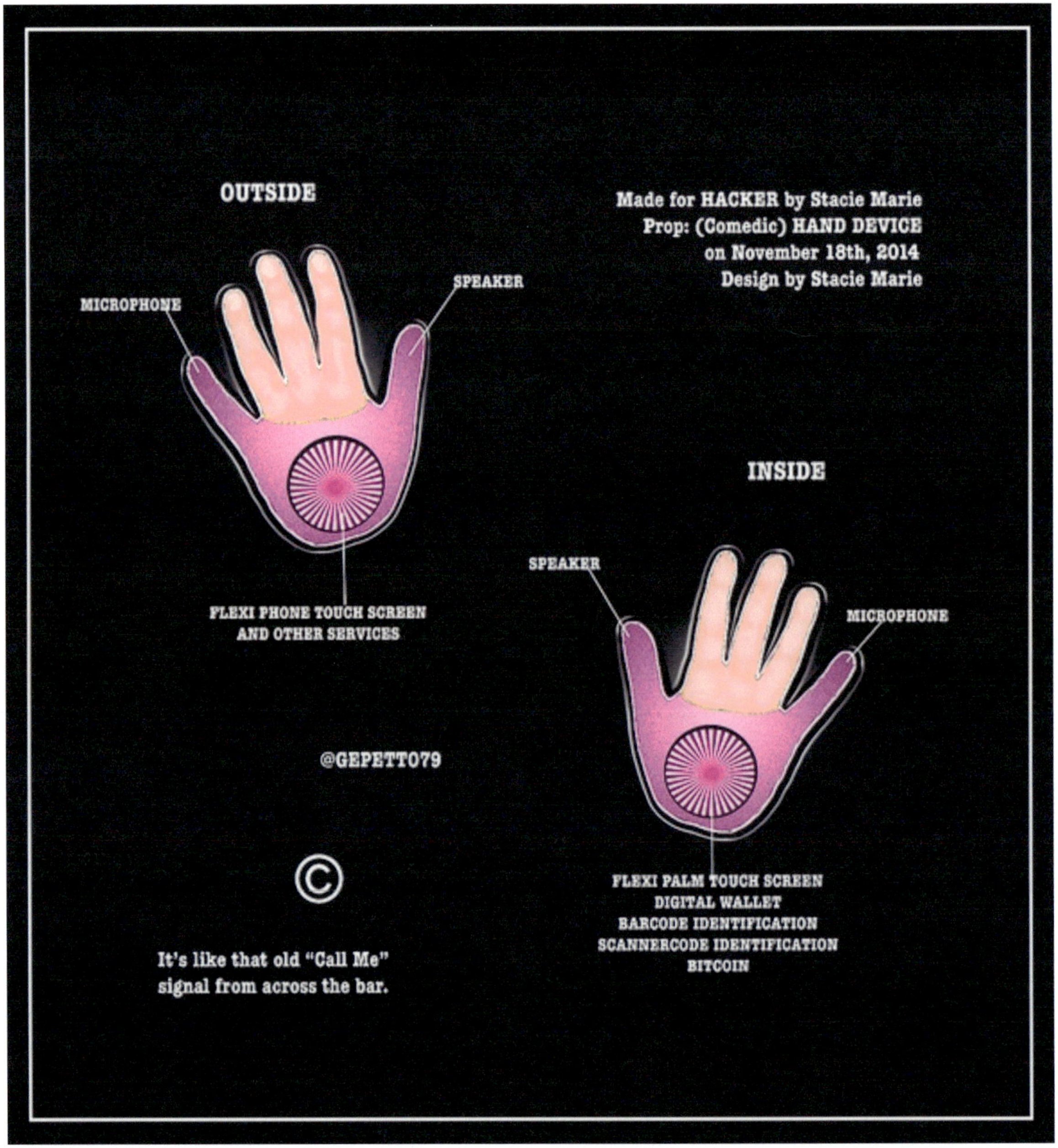
OUTSIDE
Made for HACKER by Stacie Marie
Prop: (Comedic) HAND DEVICE
on November 18th, 2014
Design by Stacie Marie
MICROPHONE
SPEAKER
FLEXI PHONE TOUCH SCREEN
AND OTHER SERVICES
INSIDE
SPEAKER
MICROPHONE
@GEPETTO79
©
FLEXI PALM TOUCH SCREEN
DIGITAL WALLET
BARCODE IDENTIFICATION
SCANNERCODE IDENTIFICATION
BITCOIN
It's like that old "Call Me"
signal from across the bar.

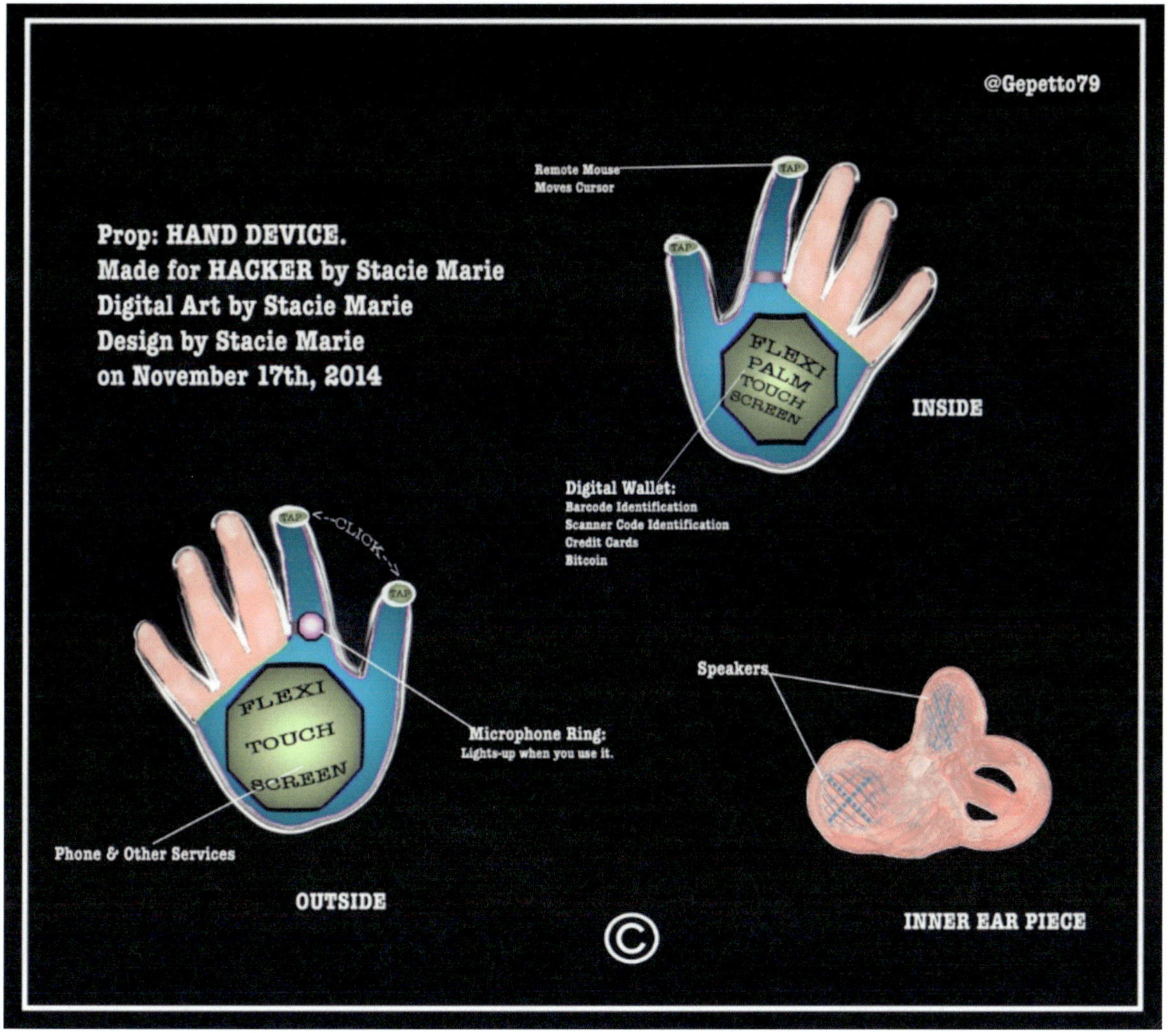
@Gepetto79
Prop: HAND DEVICE.
Made for HACKER by Stacie Marie
Digital Art by Stacie Marie
Design by Stacie Marie
on November 17th, 2014
Remote Mouse
Moves Cursor
TAP
TAP
FLEXI
PALM
TOUCH
SCREEN
INSIDE
Digital Wallet:
Barcode Identification
Scanner Code Identification
Credit Cards
Bitcoin
TAP
<--CLICK-->
TAP
FLEXI
TOUCH
SCREEN
Microphone Ring:
Lights-up when you use it.
Phone & Other Services
OUTSIDE
Speakers
INNER EAR PIECE
©

Yes
Tracing Paper + Pens
Magnetic Inlay
Fiberglass Top
Green Tube Lamps
Concept Piece:
Light Table
The Light Table (for kids)!
How it works:
Print out an image. Place it on the table.
Light it up with the green-energy light-
bulbs. Use magnets to hold tracing paper
over the images. Kids trace the images to
help them develop motor skills and dexter-
ity + they make exciting works of art.
Comes with alphabet letters, images,
special tracing paper (in various colors)
tracing cards (for special occasions).
Special printer (optional).
©

We invent a car with a windmill
attached the the roof and a small
battery to catch the reserve.
#PerpetualMotionMachine
Societal #uppers

I love you.

The end.

Made in the USA
Las Vegas, NV
04 November 2026

57425805R00021